SAR SURVIVAL

SAR Survival

SEARCH AND RESCUE FUNDAMENTALS FOR THE OUTDOORS

Jason Hunt, Ph.D.

Barefoot Prophet Media

Contents

1	An Overview of SAR	1
2	Map & Compass	16
3	Clue Awareness	26
4	Search Techniques	36
5	Finding the Lost Subject	47
6	SAR Devices	52
7	SAR Knots	60
8	SAR Response Kit	64
9	SAR Survival	71
10	SAR Documents	83

1

An Overview of SAR

Basic Search & Rescue

Basic Search and Rescue consists of three primary topics: Search, Rescue, and Survival/ Patient Support.

Search

The Search information begins with the theory and philosophy of searching for lost subjects and proceeds into the skills and resources, concluding with applying specific search tactics, including how to perform a land search.

All students of SAR should seek a chance to apply what they learn from this text regarding search theory and tactics by performing a search exercise under qualified supervision.

Rescue

At this course level, Rescue is not intended to teach you how to perform any rescue necessarily, but to educate those who will probably be involved in a rescue situation. Some of the equipment and terms used will be presented so students can identify commonly used equipment. Rescue is a highly specialized field of study with technical specifications in numerous areas ranging from rope rescue to low-angle, high angle, air, moving water, swift water, white water, wilderness, and urban naming only a few areas of specialty. At the very least, you should seek training on how to assist in packaging a subject in a Stokes basket for wilderness pack-out, as these are the most common basket types across the country among volunteer agencies.

Throughout this text, we will refer to lost people as Subjects or Patients instead of victims. A victim in a SAR environment is typically considered one who has died due to the emergency.

Survival and Support

Having a basic understanding of wilderness survival insofar as making a temporary shelter, making fire, and maintaining your own and a patient's core body temperature is critical as you may be forced to provide patient care until other members of a Search Team or perhaps EMS arrives on the scene.

SAR Students simulating patient care in the field

Your Responsibilities: Where do you fit in?

To understand where you fit in, you must first examine your motivation for serving in search and rescue. If you are in it for any other reason than to help others, then your career will be short-lived. The one reason we serve in SAR is summed up in the motto "That others may live" - we do it so others may survive, we seek to reduce suffering and pain for the subject of our toils.

We also spend our own money, time and expend our efforts to save others that will, in most cases, not give us thanks or know our name. Search and Rescue rarely gains us any recognition.

As far as a career choice:

- There is very little to no money in SAR
- There is rarely any acclaim
- SAR work is tiresome, uncomfortable, dangerous, and rarely required when we have time to do it
- Rarely is there even a "thank you" in SAR

Sounds good, right? All this necessitates that an individual is focused and confident. In light of this, there is no more sweet satisfaction in anything than in knowing that you have risked your safety and possibly your life to help another in their most significant time of need. (John 15:13)

Laws about Search & Rescue

In all states, the local division of emergency management is required to have:

- A state SAR plan
- A state SAR Coordinator
- Resources for Searches
- Date concerning searches

In addition, most counties in every state are required to:

- Have a county SAR plan
- Have a county SAR Coordinator(s)
- Designate agencies in the county to conduct searches and rescues in the county, be they independent SAR Teams or local Fire Departments.

The county emergency management director or highest elected official often organizes these county activities, such as a County Judge Executive. Please check your state and county laws for specific SAR Laws and Team development information.

Good Samaritan Laws generally speaking only apply to those not involved as part of an organized rescue squad or other agency. Thus, you as an individual, voluntarily rendering aid to another, would be protected to some measure by such laws. However, if you are actively engaged in a SAR deployment and are part of the team, you would not be covered and expected to help the level your training dictates.

The Importance of Ground Searchers

Of the types of resources utilized for SAR efforts such as dogs, aircraft, and satellite images, nothing takes the place of a "ground-pounder," the lone search technician that provides actual intelligence on clues, terrain, and information that can prove whether a subject is actually in the area or not.

Crucial Concepts

Search is a true emergency!

- The Subject may need emergency care
- The Subject may need protection from self or the environment
- The Subject may be responsive for hours or even days
- An urgent response lessens search difficulty

Search is a classic mystery that requires looking for clues that lead to a lost subject. Concentrate on aspects that are important to success under the control of the Search Manager and learn enough to know when the subject leaves a search area. The Search Manager is the information manager on a SAR Team.

SAR Professionals

SAR workers need to understand precisely what is expected by their managers and leaders; the acronym 'PHACKS' was devised to convey these expectations.

P: Professional/ Proficient

- Perform with expert correctness, be adept
- Do what you do well in a professional manner (look and act the part)

H: Humility

- Show awareness of your shortcomings, do not be prideful
- Don't talk about doing a good job, do your best and let others be the judge

A: Able

- Be capable of performing, both physically and mentally

C: Competent

- Be appropriately trained and qualified
- If rescue is required, be qualified to perform it; if search, then be equally as qualified

K: Knowledge

- Be familiar and aware through experience and study
- Neither experience nor study exclusive of one another can be depended upon in SAR education

S: Solicitous

- Be concerned, attentive, and eager

- Are you the first to get going when there is something to do, or do you wait for another to take the lead? Be the leader.

Freelancers

Freelancers have some SAR training or interest yet have no formal team affiliation or appropriate agency under which to operate. They can be anyone from the eager bystander, team member from "another" department to those that wait for things that catch their ear over the police scanner. Freelancing is not at all well received in the Search and Rescue world. Showing up to a SAR emergency without being requested could lead to you being removed or arrested by law enforcement.

If your plan includes responding to SAR emergencies, it is best to be formally affiliated with an existing SAR Team in your county. If your county does not have a SAR Team, just find out the agency responsible for conducting search and rescue within the county and follow up there. In most cases, that agency will be the local volunteer fire department.

Authors Experience:

When I became interested in Search and Rescue, the nearest team I could locate was over an hour away. I still joined and completed my basic training with them but never responded to an incident. It wasn't until a year later that I could track down the local agency responsible for SAR activities, the Fire Department. When I joined the fire department, I had no interest in fighting fires but merely serving on the search and rescue team. But to join, it was required that I had at least Basic Fire training. As it turned out, I thoroughly enjoyed the fire training and became an asset to my department, and to top it off, I was the only person in the department that had any actual SAR experience. This dual qualification enabled me to take advantage of all sorts of free courses that have helped me in many areas of my career in and out of emergency services.

Responding to a Search

The Initial Report should detail

- The exact location of the search
- The time you are to arrive on the scene
- The location of the staging area and on-scene contact person
- Specific instructions as to the location of the staging area
- What is expected of responding resources (job duty)
- What Topo maps constitute the search area
- Info related to Lost Subject
- Radio Frequencies in use or radio provided
- Special equipment needed (ATV, Dogs, Generators, etc.)

If not provided, ask these questions BEFORE you go to respond.

Upon Arrival at Scene

- Get instructions on what you are to be doing
- Report to your designated leader/manager
- Sign any required forms
- Identify your team and the number of people in that team
- Get ID or Accountability Tags

Before Deploying

- Find out deployment information
- Know your intended search area
- Where your staging area is located
- When deployment begins
- Check necessary equipment and maps before entering the search area

Helpful Classes provided by state and local emergency services agencies:

Basic Search & Rescue (Also called FunSar or BSAR)
Intermediate Search
Mantracking
Urban Search and Rescue
Lost Person Behavior
Search Management
Rope Rescue
High Angle Rope Rescue
Wilderness & Remote First Aid

Moving and Swift Water Rescue

Levels of advancement are typically recognized as:

- Awareness Level 8-20hr of training
- Operations Level 21-39hr of training
- Technician Level 40+ hr of training

The National Association for Search and Rescue, NASAR, also offers courses and certification levels for a fee and is widely recognized as the national standard for SAR Professionals. This text and a hands-on course will enable you to challenge the NASAR SARTech courses for national qualification if desired. Find out more at www.nasar.org

Briefing

The briefing is one of the most critical activities on a search. A poor briefing can result in poor search implementation, non-searched segments of land, misuse and destruction of clues, and ultimately failure to find the lost subject. The briefing can make or break a search operation; everyone must be briefed orally or written.

A briefing provides

- Situation Status
- Search Objectives
- Search Strategy
- Tactical Assignments

Who coordinates a briefing?

The planning chief or delegated person in the Planning Section of an extensive search operation.

Who briefs, and who gets briefed?

Everyone is involved in the search process!

The Planning Chief

- Briefs Overhead Team (Command)
- Briefs Team Leaders, Agency Liaisons, and more minor ops

The Operations Chief

- Briefs supervisors/group leaders in more extensive settings
- Briefs Team Leaders in medium-sized operations
- Briefs Team members

What information needs to be presented?

The following information should be given to:

Team Leaders:

- Situational Status (Objectives/Strategies)

- Subject Information- all info that will aid searchers in locating the lost subject and finding clues such as
- Name
- Physical Description
- Clothing and Equipment
- Physical Condition
- Mental Condition
- Behavior Traits
- Circumstances Causing Search
- Vital Concerns (Health/Medical Needs)
- Clue Considerations such as
- Sole pattern of footwear
- Safety equipment
- Food and Water
- Recording Equipment
- Special Equipment (ATV, Gun, Lights, etc.)
- Communication Details (Codes, etc.)
- Reporting Details (When, Where, How, etc.)
- How long teams will be deployed
- Who are relatives or friends in relation to the search area
- Media Procedures- who, what info is shared, who is the media liaison
- Tactical Assignments with explicit searching instructions for the team such as a Specific area to search
- Search configurations and spacing
- Expected probability of detection (percentage of the area searched)
- Clue marking procedures
- Adjacent teams/ known hazards

- Have others already searched the area
- When to start/stop, what to do if the subject is found alive/dead/injured, and instructions on protecting the scene.

Team Leaders will then assign individualized objectives to the search that will emphasize the following information:

Medical Plan/ Rescue Evaluation Plan

- Team members will know who is in charge of what
- Debriefing instructions (when, where, and who to report to and what to report)
- Safety instructions- safety hazards, traps, snakes, dog hazards, terrain issues
- If possible, an estimate of when the team can go home or rest
- Procedures if team members get injured

Briefings should be time-limited
Typically no longer than 30 minutes to:

- Maintain and improve the morale of personnel
- Maintain management credibility in the eyes of subordinates
- Ensure timely response to the field and maximize time for task accomplishments
- Keep it simple and to the point!

Key briefing techniques

- In every case possible, distribute written briefing statements, task statements, diagrams, maps, photos, sketches, etc., to convey information.
- Photos of the subject
- Briefing statement photocopied
- Sketch of the sole pattern
- Map to show assigned area and search details

Section 1 Quiz

This quiz is to aid in your retention of information provided thus far. You will find quizzes and exercises throughout this text.

1. What are the four things that local emergency management divisions are required to have in all states?
2. Define PHACKS:
3. Search is not an emergency. True or False?
4. What are the four crucial elements for any search operation to consider seriously?
5. Freelancing is encouraged to aid search teams with a few members with interested, trained parties to assist them. True or False?
6. Who coordinates a briefing in a more extensive search operation?

7. Name four things that would be considered subject information.
8. How long should the average briefing take to conduct?
9. A Team Leader may be anyone willing to take on the role. True or False?
10. Rescue requires little specialized training. True or False?

2

Map & Compass

Compass Reading

The instruction provided here is intended to acquaint you with some general terms and the essential parts.

Magnetic Declination is the angle between True North and Magnetic North. Maps are commonly based on True North, while the needle on a compass is drawn to Magnetic North, assuming it's not affected by metal objects on your person such as heavy watches, buttons, etc., or overhead power lines.

For example, Magnetic North has a five-degree variance from True North at my property, which equals about fifteen feet; In comparison, the variance is as much as sixteen degrees or roughly 48 feet in California. This amount may seem small over short distances, but the gaps between the two widen significantly over five hundred or more meters. When a compass is used with a Topographical Map (Topo), an adjustment should be made for declination to compensate for the variance between true north and magnetic north. This adjustment will often be noted on the map itself. This becomes especially important if there is considerable variance in your region- you must adjust the compass or compensate for the variance in your readings- otherwise, your direction of travel may be in error. While this may not be a big deal in areas such as mine in Kentucky, where I'm only off by as much as fifteen feet, it will make a significant difference in areas further out west where the variance can be anywhere from fifty to one hundred feet in as little as one hundred meters. That can mean the difference between finding your lost subject and not.

Many compasses you purchase today have already been adjusted for zero declination, so they may be adjusted for the region in which they are used. You can also look for a compass that has already been changed or read the directions for yourself and make the adjustment as needed.

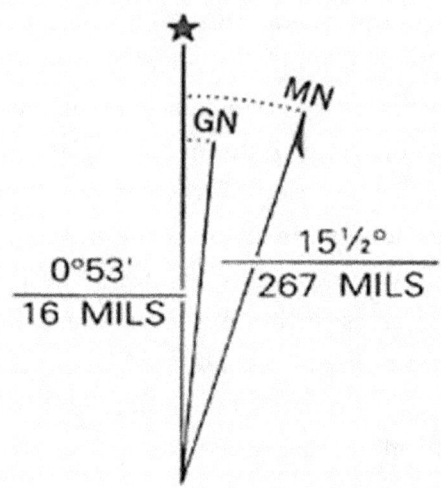

UTM GRID AND 1994 MAGNETIC NORTH DECLINATION AT CENTER OF SHEET

Declination as shown on a Topo Map. The Star represents True North, GN: Grid North, and MN: Magnetic North. GN is only 53 feet off True North, while MN is over 15 degrees off.

An example of Quick Reaction or Hasty Team SAR Navigation Kit may include a Topo Map, Compass with Mirror, Pace Beads and essential survival tools to care for a lost subject or the rescuer.

The author's quick reaction SAR Kit

Maps

Topographical maps are like aerial photographs of a terrain taken at a high altitude and reduced to a scale. A map identifies its scale as 1:24,000 – one inch equals 2000 sq. feet. One unit on a map equals 24,000 units on the ground. You must note the scale of the maps you choose to utilize and learn what distances are indicated on those maps. It is recommended that teams use a 1:24,000; 7.5 grid topo map when available for SAR purposes.

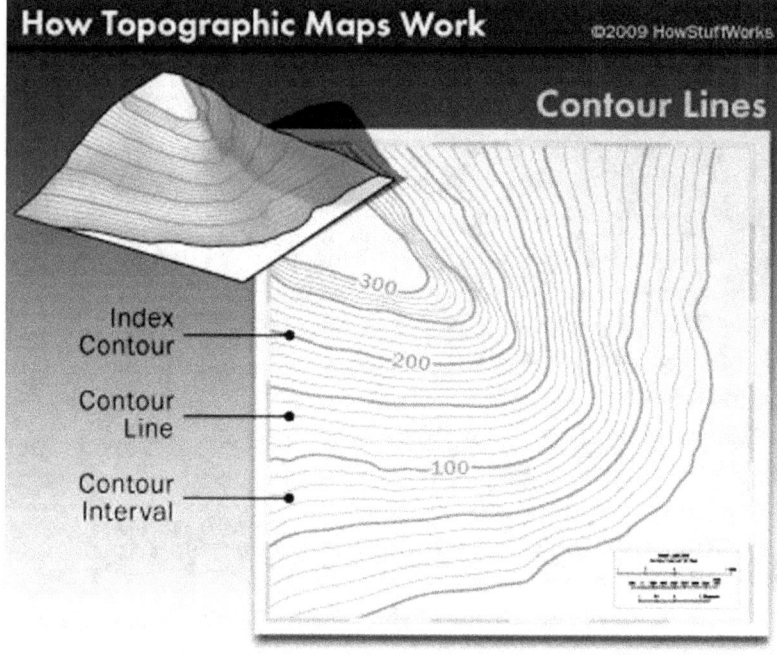

There are many types of topo maps available, and they are generally color-coded and depict the shape of the earth's surface using contour lines. Contour lines are the imaginary lines that follow the terrain's surface at a constant elevation above and below sea level. The use of color helps distinguish various features.

Map Colors

- Black: side roads and buildings
- Blue: Water of all types (lake, stream, etc.)
- Brown: Contour, elevation, and boundary lines
- Green: Woodland Cover, shrubs, etc
- Red: Main roads and public land systems

- Purple: features added from aerial photographs

When searchers use topo maps, there is one principle to keep in mind. Most USGS maps were printed years ago, and many have not been updated since the 1970s or even earlier. When using these maps, consider recent changes to the area, such as new construction within the last decade or buildings that may no longer exist in addition to new roads and recent developments, and even artificial waterways. Therefore, it's vitally important to have someone with current knowledge of the area, the terrain, and, if possible, to be from the area to be searched while initiating a search strategy.

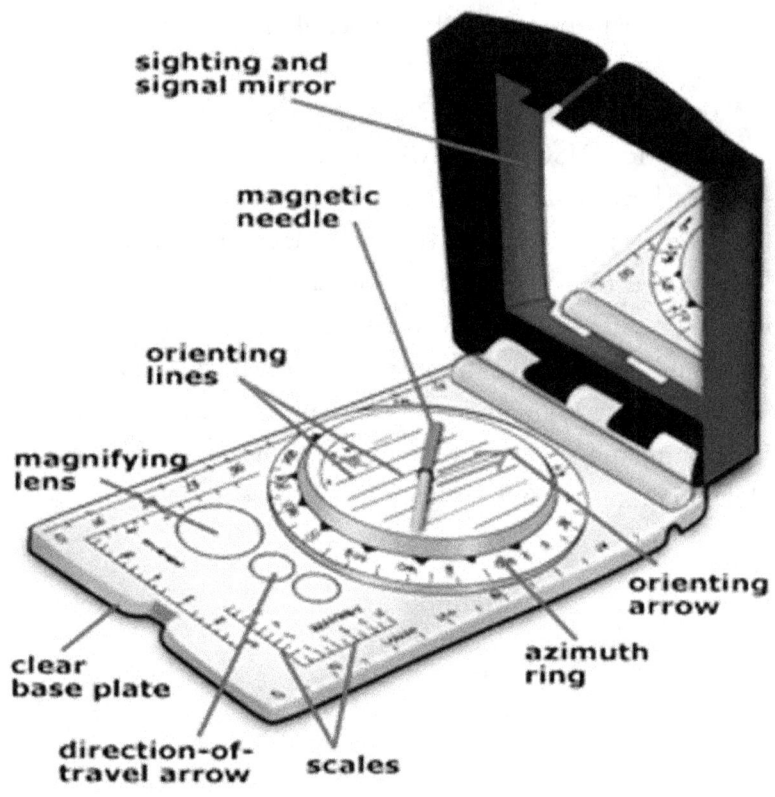

Compass Nomenclature

Map and Compass Exercise

Three-Leg Compass Walk
Start by having each student mark their location with a pencil or a small stick and set their compasses to the north (360 degrees). Once north has been established, they sight down the direction of the travel arrow on the compass and pick out a landmark in the background. The landmark can be a tree, baseball

backstop, telephone pole, or another stationary object. The student should then step out one hundred paces- a pace is counted each time the left foot hits the ground as you walk- everyone stops after 100 paces.

The student is then directed to set their compasses to 120 degrees, walk one hundred paces, adjust the compass to 240 and walk another one hundred paces. At this point, they should have walked a triangle and ended up reasonably close to where they started.

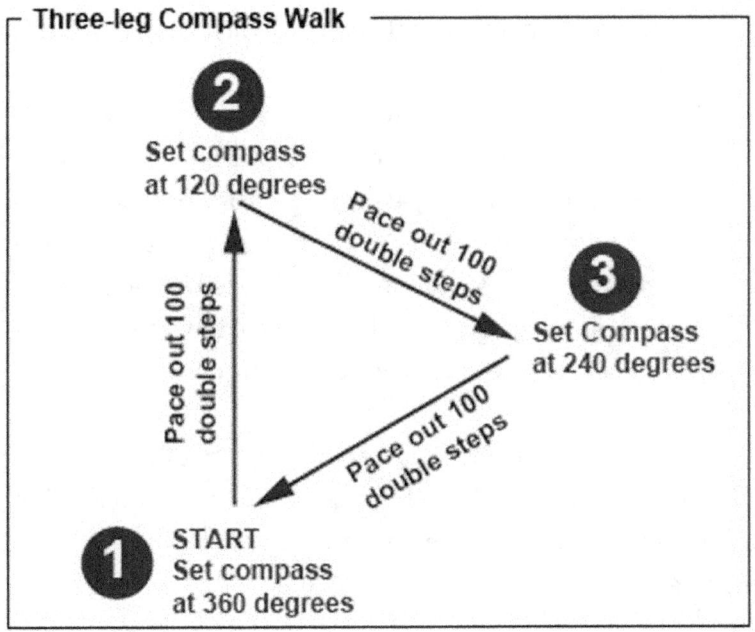

To Use Map and Compass Together

1. Check the map's legend for declination and adjust your compass as or if needed.
2. On the map, draw a straight line between your location and where you want to go.
3. Place the compass down on the map, line up the index lines with the directional line you drew on the map.
4. Rotate the bezel ring until the orienteering arrow lines up with the meridians (north/south lines) on the map and points to the Northside of the map. Note- the needle can point any direction at this point; you'll adjust it after you have a bearing.
5. The degree indicated on the bezel ring in line with the direction you want to go your map bearing or azimuth.
6. Now, hold your compass level in front of your body and rotate your body until the North arrow and Orienteering arrow line up. We call this "keeping red in the shed."
7. Now, move in the direction indicated by the direction of the travel arrow until you reach your destination.

The map bearing and field bearing are two different critters. The Map Bearing, explained above, is used with a map. When a map is not available, a field bearing is taken. To utilize a field bearing:

1. Look in the direction you wish to travel and note a large, stationary object.

2. Holding the compass in front of you, rotate the bezel ring until red is in the shed, the number now pointing in the direction you wish to travel your field bearing. Write this down.
3. Look at the terrain; is there anything that may take you off courses, such as steep inclines or drops, thick brush, or body of water. If you reach such an obstacle, you may need to aim off to go around the obstacle before getting back on course, requiring an understanding of pace counting.
4. If no obstacles are detected, double-check your field bearing and move in that direction of travel.

It is highly recommended that a complete, hands-on course in basic navigation be completed. Most survival instructors, Rescue Squads, and Boy and Girl Scout groups offer training. Parks and Recreation agencies and schools also typically have someone qualified to assist with compass usage.

3

Clue Awareness

Clues are facts, objects, information, or evidence that helps solve a mystery or problem. Search team members must search for clues in addition to the lost subject. This is because there are always more clues than subjects; the detection of clues reduces the search area, and the information gained from clues may give you information about the location of the lost subject.

Be clue conscious!

- Clue seeking is a significant job of the field searcher
- Clues can be found, discovered, stumbled upon, etc.
- Good clue seeking is learned and practiced frequently to develop and maintain skill level
- Opinions must be formed based solely on the information available; not by gathering information to support an opinion

Clue seeking assists us in our reasoning of the problem (finding lost subjects) by gathering all the facts and information possible. Significant clues may provide the basis for effective tactics and actions in the field.

Clue Generators (The Lost Subject)

Virtually every person that passes through an area leaves evidence of their passing. A common problem is not the lack of clues but determining which of many possible clues relate to the search and are valuable. A detailed subject profile enables searchers and managers to connect a particular clue to the subject or not.

The 6 Elements of Clue Oriented Search Theory

1. The Subject or Clue Generators
2. The Clues themselves
3. The search area where clues are located
4. The searchers or clue seekers
5. Chronological order of events as they relate to the search.'
6. Search methods and tactics used to locate clues

There are four messages that a lost subject can convey that a searcher must be able to detect during a search event:

1. Present Location
2. Previous Location
3. Destination

4. Total lack of clues

Have a plan to deal with any clues found. How Incident Command deals with clues found will be presented to searchers in the briefing.

4 Clue Categories

1. Physical- footprints, cigarette butts, broken vegetation, discarded items, overturned rocks
2. Recorded- summit log, trail register, boat or ATV rental
3. People- witness, family, friends, people in the search area
4. Events- flashing lights, whistle, yell, etc.

The Importance of clues

- It's vital to search for both clues and the lost subject
- It's crucial to look side to side, down to up, and even behind you to gain a different perspective of the search terrain
- Another way to attempt to locate the subject is to blow a whistle or call their name and wait to listen for some sort of response. This must be coordinated through command, so EVERYONE on the team goes quiet to listen.

If the clue can be related to the lost subject, you may be able to discover:

- Direction of Travel

- Physical sign or mental condition
- LKP- Last Known Position

Signs

If the clue is in cold weather and includes discarded clothing or equipment, it may indicate that the subject suffers from hypothermia. If the clue is found in the summer and contains empty water bottles and clothing, the issue may be suffering from hyperthermia and dehydration.

A series of clues can be linked together to tell a story, determine a direction or possible place of refuge. An indication such as a fire or shelter can indicate the subjects' behavior or knowledge of survival techniques. The subject may not be far off if the fire is still warm. It's essential for searchers to continually look for signs that the lost subject is trying to communicate to them.

Signs that are intentionally recognized in groups of three are:

- Gunshots
- Fire
- Whistles
- Stacked Rocks

The following may not be in groups of three:

- Pyrotechnics (Flares, Rockets, smoke generators, etc.)
- Direction of travel arrows

- Subjects with knowledge of survival techniques may utilize ground to air symbols or emergency codes such as S.O.S. Of three X's on the ground.
- Some subjects may leave behind clothing or notes along their path of travel

If the clues are correctly reported to the Incident Command, then the command staff can work with the relatives of the lost subject to determine if the clues are those left by the missing subject; the subject's actual travel or condition may be determined, and a new search area assigned.

SAR Survival students making a Smoke Generator

Clue Preservation

You may:

- Photograph clues
- Sketch clues in relation to other clues and terrain
- May use a cell phone to photograph clues to transmit to command
- Make notes and description of clue in note pad

Before searching for clues on private property, you must first obtain permission from the landowner. This is typically done by Incident Command, but not always. Be courteous of the landowner's property- do not disturb or destroy any property. Be aware of any livestock and if a gate is closed, make sure that you close it back as you go through it; if one is open, make sure it remains open.

Protecting Clues

Consider how clues should be protected:

- Does the area need to be cordoned off?
- Do plaster casts need to be made (of tracks)?
- Does the area need to be guarded?

If cordoning off or plaster casts are required, a technician will be sent to handle these matters; in the meantime, do not smoke or leave any other unrelated clues at the scene, which could complicate the handling of evidence. If you find a dead body or other human remains, do not move the body or move anything in the area. Many states require a call to law enforcement and the county coroner when remains are found so they may report to the scene. Someone, typically the one that sees the remains, must remain on scene until law enforcement or the coroner arrives.

If a downed aircraft is found, its location must be reported to local law enforcement, The Air Force Rescue Coordination Center, the 911 Call Center (Typically your State Police Post), and the National Transportation Safety Board (NTSB).

If the wreckage of a bus or other commercial vehicle is found, local law enforcement and NTSB must be notified. Local law enforcement and State Police must be notified if a wrecked car or other clues are found.

DO NOT TOUCH THE CLUE OR REMOVE IT UNLESS DIRECTED TO DO SO BY INCIDENT COMMAND!

Sign Cutting

When a clue is located, it is usually wise to circle the clue to see if a direction of travel can be determined. For instance, if footprints or ground disturbances are found within a 360-degree area around a clue, you may be able to decide on the lost subject's direction of travel.

If no clues are found, and the area has been thoroughly searched, it may indicate that the subject has not been or is no longer in the area. This is also helpful information as it tells search managers where to look next.

Marijuana Fields, Meth Labs, and Booby Traps

Finding pot, stills, meth labs, and other drug and alcohol-related areas in the wilderness is more common than you might think, especially in underdeveloped areas with poor economic conditions. Here are some indications to be aware of...

Indicators of Illegal use:

- Empty Milk Jugs
- Fertilizer Bags
- Propane Cylinders
- as Jugs
- Sudafed or Pseudoephedrine boxes
- Strong odors of ammonia
- Poisonous snakes tied to trees or staked to the ground
- Rat traps nailed to frees with shotgun shells in them
- Trip Wires
- Perimeter Alarms with cans hanging on them
- Any other object that is drug or alcohol-related and is out of place in the woods

If you find evidence of drugs or alcohol in the area, you should:

- Leave immediately and notify Incident Command so it can be reported to Law Enforcement
- Be discreet; you don't know who may be listening or watching. It's common for family members to be around search scenes; it may be their stuff or someone they know

- Do not use radio communication as the method of contact with command; if you find something to state, you are terminating search in that area and will report in person upon return

Evaluating Search Effectiveness

- One of the most important things to practice is clue seeking
- One of the most critical questions for a searcher to be able to answer is "What is my probability of detection (POD)?" or what percentage of your area have you searched
- If you are unable to locate clues in an area or have a low POD, this may mean the area will need to be searched again, or perhaps the search area should be widened based on the information you provide
- looking for clues is a skill that you as a searcher must practice regularly

Terrain Evaluation in the Search Area

- Terrain may indicate where and how a subject may travel and how long they may survive
- Look for forks in the road or trail which may confuse a lost subject
- What sites or sounds may have been witnessed or heard from the PLS (Point Last Seen) or LKP (Last Known Position) that may have attracted the subject toward them?

- Are there natural barriers that limit search or travel? (Cliffs, Rivers, etc.)
- Are there natural routes that offer less resistance to travel? (Game trails, ATV trails, etc.)

This is vital information that should be passed to Command during debriefing.

Section 3 Quiz

1. Define LKP:
2. Define POD:
3. Define PLS:
4. Name 5 Indicators of a drug or alcohol area in the woods.
5. Name four ways of preserving clues.
6. Give 3 Examples of a sign.
7. Name 3 of the six elements to clue oriented search theory
8. If a clue is related to your lost subject, you may determine?
9. How can clues be found?
10. What are the four categories of clues?

4

Search Techniques

There are three methods for determining how or where you should search.

1. **Theoretical Search**- this area is determined by locating LKP or PLS.

   - Takes into account elapsed time from the time the person went missing until the search area is determined
   - Figures out how far the lost subject could have traveled
   - Establishes a maximum distance within a 360-degree radius or PLS or LKP

Example: If the lost subject traveled at three miles per hour and has been missing for two hours, that is a radius of 6 miles in every direction from the PLS or LKP. Over ten hours, it would take a small-town population to cover that much terrain at an 80% POD!

1. **Statistical Method**- this area is based upon past searches using information from subjects of similar age and condition.

- It limits the search areas based on age, distance traveled, and condition of past lost subjects of similar age and types that have been found
- While this method is an improvement over the theoretical method, experience search managers also utilize deductive reasoning to determine which areas to search first

1. **Deductive Reasoning** – the search area will be limited or expanded based on the lost subjects' travel barriers.

- Rivers, lakes, cliffs, or other areas where subjects are not likely to go are balanced against things that may attract the subject to alter their paths for more accessible travel or places of interest
- If someone went hunting, where's the best hunting spot?
- If someone went fishing, where's the best fishing spot?
- If someone went hiking, where are the best trails?

Patterns

The initiation of search tactics takes place simultaneously as establishing the search area. Search tactics are categorized as:

- Passive- Confinement and Attraction; you make the subject come to you
- Active- You find the subject

Passive Techniques may include:

- Smoke
- Lights or Fire
- Sirens
- Whistles
- Horns
- Loud Speakers

Confinement Techniques may include:

- Point control at road and trail intersections
- People coming out of an active public search area are asked about clues they may have found
- People going into the search area are asked to keep an eye out for clues and evidence of the lost subject
- Running roads and trails with ATV's, mountain bikes, and on foot maintains a workable perimeter and search vein activity

The 4 Active Types of Search

Type 1: Rapid Response to areas of a high probability of detection by immediately available resources.

Criteria: Speed

Considerations:

- Assumption is you are looking for a responsive subject
- Provides an immediate show of effort
- Can help determine search area by gathering information or locating clues
- Clue consciousness is critical
- Often results in where not to search further
- Pre-planning is crucial for practical use and deployment of resources

Techniques:

- Investigation (Personal, Physical Effort)
- thorough checks of LKP
- Follow known or suspected route
- Running Trails
- Perimeter Check
- Sign Cutting
- Road Patrol
- Check area attractions
- Check area hazards
- Check drainage's
- Run Ridge Tops

Most Effective Resources:

- Investigators
- Trained Hasty Teams (Rapid Response Searchers)
- Mantrackers
- Dogs (Air Scent, Trackers)

- Aircraft (Unless the area is heavily wooded)
- Every other mobile, trained resource: ATVs, Horse Teams, Snow Mobiles, etc.

Type 2: A fast systematic search or a high probability segment of search area using techniques that produce high detection probabilities per searcher hour of effort.

Criteria: Efficiency

Considerations:

- Often employed after Type 1 efforts in some segments of the search area, especially if Type 1 found clues
- May be the initial search tactic used in search segments, particularly heavily vegetated areas
- Should be used when subject responsiveness is still expected to be high
- Type 2 efforts are often effective in locating clues

Techniques:

- Used primarily in specifically defined search areas
- Used to follow up in a segment where a clue has been found
- Uses an open grid with wide spacing between searchers
- Search routes are often followed using compass bearings

Most Effective Resources:

- Investigators
- Clue conscious teams
- Dogs (Air Scent, Tracking)
- Mantrackers
- Sign Cutters
- Aircraft
- Grid Teams

Type 3: A slower, highly systematic search using specific techniques.

Criteria: Thoroughness

Techniques:

- Closed Grid or Sweep Search

Most Effective Resources:

- Trained Grid Searchers

Type 4: A slow systematic (fingertip) search similar to an archaeological dig. Also known as an Evidence Gathering Search. Once a clue is found, a string is run on a datum and baseline.

Criteria: Extremely Thorough

Considerations:

- If not done correctly, evidence could be destroyed as well as the crime scene
- Requires a lot of manpower
- Very time consuming

Techniques:

- Search teams wear protective clothing and knee pads and start searching in grid squares, working their way forward
- Once an object is located, the position is triangulated back to the datum point
- Police want all evidence photographed, tagged, and bagged by their exhibit officer

Sweep Searches

Martin Colwell of Canada has developed different types of sweep searches. The type of search depends on the assumption of the subjects' condition, clothing, ability to communicate, and mental status. The most frequent search used for a lost subject who can communicate is the sound sweep.

Sound Sweep: For persons thought to be alive and responsive, a loudspeaker broadcasts blasts over a timed interval as a beacon to draw subjects toward it. Five seconds of radio silence follows after each discharge to listen for sounds from the subject.

Standard Sweep: For adults and children wearing everyday colored clothing (neutral visibility).

High Visibility Sweep: For persons known to be wearing high visibility or brightly colored clothing, typically in more open terrain.

Low Visibility Sweep: For those known to be wearing earth tone clothing

Body Sweep: For those presumed dead.

Sweep Searchers are more common in Canada and may be utilized by specific teams trained by Canadian sources from time to time. It's becoming less common in the United States but is still found along with border states to Canada.

Day and Night Searching

Your vision is different in the daytime than at night, and that difference can profoundly affect your search capabilities.

How vision reacts to daylight:

- Pupils constrict
- Colors and fine details are seen
- Visual acuity is at an optimal level
- Images are perceived towards the center of the field of vision

- Scanning requires concentration, a set routine should be used (up and down, side to side)

How vision reacts to darkness:

- Pupils dilate to let more light in, meaning you will need to use peripheral vision at night, and it will make your eyes longer to adjust to the darkness. Your vision will be improved as you look around instead of focusing on an object- you'll look around to perceive an outline or catch movement
- Scanning requires concentration, use the same techniques as day scanning, but view the object off-center in your field of vision instead of straight on
- It takes at least 20 minutes for your eyes to adjust to the darkness
- Avoid night blindness by not depending on the beams of bright flashlights and by not shining lights into the eyes of other team members; also, do not look at the flashing lights of response vehicles
- Look for shapes, shadows, and contrasting movements

Night Vision Goggles

Night vision goggles (NVG) are light amplification devices that do not work in total darkness. They are merely extra weight to pack in the deep wilderness with a thick canopy or in a cave. Additionally, they burn out when other searchers introduce reasonable light. Less expensive civilian models are becoming more popular and have their merits, but generally speaking, they're too costly and do not perform well enough for standard SAR use.

Thermal Imaging Devices

Thermal imaging devices detect heat sources and do not rely on light sources. They also work better for SAR applications than most NVG's available to civilians. That said, the light emitted from the screens does nothing to help searchers' eyes in the night, and it can sometimes be difficult to distinguish between a warm body and a warm, wet pile of leaves. Most Fire Departments carry handheld FLIR (Forward Looking Infrared) devices on their apparatus, so take advantage of them should you feel the need if they become available.

Hazards of Searching at Night

- Falling off Cliffs
- Falling into holes or dry creek beds
- Eyes can become injured from tree limbs
- Bears, Snakes, Feral Hogs, etc.
- Bad Weather conditions
- Trip Hazards (Rocks, Logs, Vines, etc.)
- Becoming lost or disoriented yourself in unknown lands
- Fear of the dark- becoming panic-stricken
- Being unprepared with improper field gear

Preparing Rescue Team during a Wilderness Responder Class

Section 4 Quiz

1. Name the primary differences between the four types of active searches.
2. Sweep searches are becoming more common in the United States. True or False?
3. Name five hazards of searching at night.
4. Night vision works best in total blackout environments. True or False?
5. FLIR systems work well in wooded areas. True or False?

5

Finding the Lost Subject

When you find a lost subject, Notify Incident Command!

Use the Acronym L.A.S.T. When searching for a lost subject:

- L- Locate: Notify command of the search team and the location of the subject
- A- Assess: Assess the condition of the subject and administer first aid if needed
- S- Stabilize: Stabilize and secure the subject for extrication
- T-Transport: Transport the subject/ patient to safety

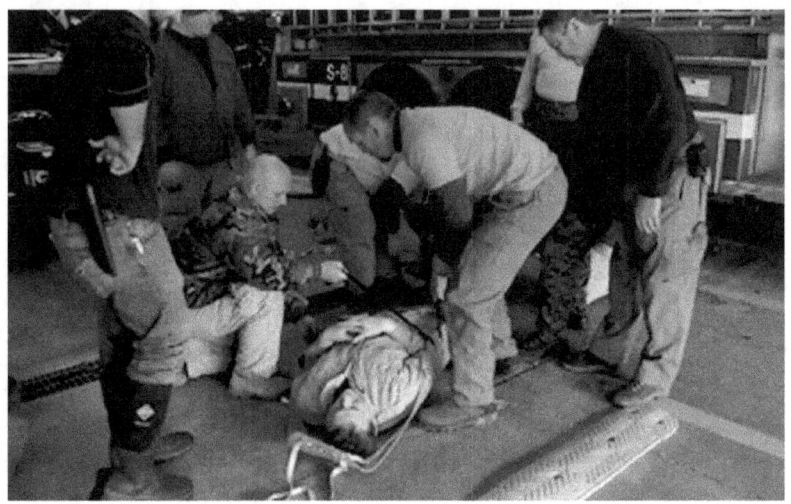

Students learning to package a patient for transport

Incident Command will assess the type of extrication and transport of the subject based upon the information you as a searcher/rescuer provide. Command will also call in resources as needed for transportation of the subject while determining whether or not the subject is to be transported by:

- Air (Helicopter)
- Ground (Hand Carry, ATV, 4x4)
- Boat

Remember:

- Who is responsible for the expenses relating to the transport of the subject?
- What resources are available? What you want may not be readily available.

- Consider the time it will take to get resources to your area for use in extrication. Searchers will have to remain with the subject until additional resources arrive or transport becomes available unless the subject can walk out.

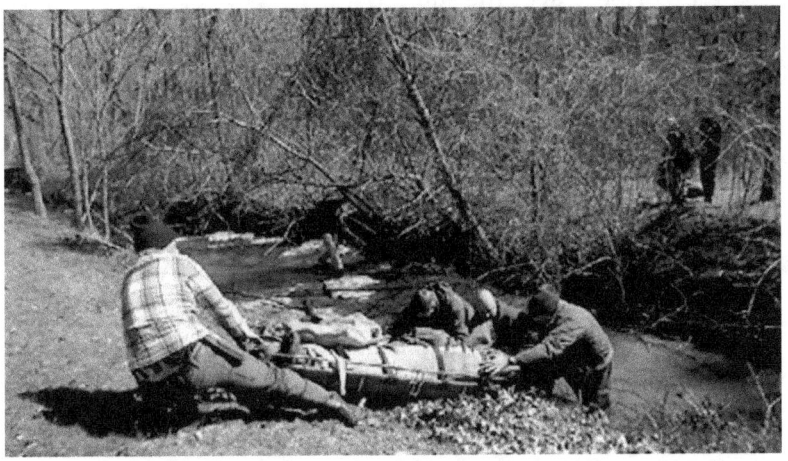

Subject being hand carried from wilderness area

If the lost subject is found deceased

- Notify Command with a code that has been determined during the brief such as Code Black, which would mean deceased subject.
- Command will notify law enforcement and coroner and will recall search teams not in the area of the deceased
- Keep lines of communication open, no unnecessary radio chatter
- Remain a safe distance away from the subject while not leaving the area
- Do not touch or disturb the scene

- Secure the area and only allow law enforcement to enter the scene
- Remain at the scene until relieved by law enforcement, the coroner, or until recalled by Command.
- Be respectful of the subject.

Debriefing

Useful information conveyed in the debriefing is essential for an effective search. Use any means to get what you want to say about the area searched (sketches, maps, reports, notes, etc.). After searching an area, searchers should report to pre-designated individuals within Incident Command for debriefing.

Debriefing should include

- Search area covered/not covered
- Probability of Detection
- Suggestions for next area of search
- Lost or damaged equipment
- Injuries
- Hazards found in the area
- What were hazards
- Hypo/Hyperthermia
- Mechanical Injuries
- Animals/ Insects
- Other
- Should be done in writing

- Should be conducted one team at a time when coming in from the field

Proper debriefing procedures provide Command with the tools to make proper planning decisions for the net shift in the search incident.

Estimating Probability of Detection

If there were ten clues found in a search area you were assigned, how many would you have found? 2= 20%, 4= 40%, etc. Typically, there are hundreds of clues within a designated area; the skill lies in learning to identify them.

Section 5 Quiz

1. Define the acronym L.A.S.T.
2. What is the first thing you do when locating a dead body?
3. List the eight things a proper debrief should include.

6

SAR Devices

We will now discuss some of the tools used within Search and Rescue.

Ropes

Ropes are an essential part of the SAR Teams' equipment. There are four types of rope common to SAR usage.

Laid Rope: Laid Rope has fiber bundles twisted around one another.

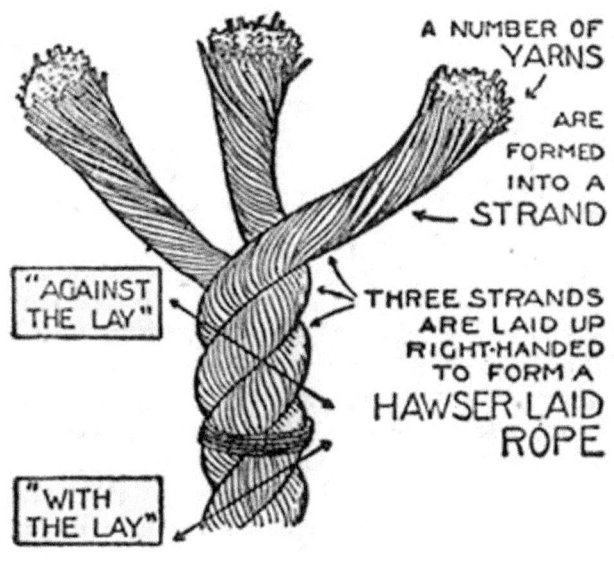

Mariner's Laid Rope

Kern Mantle: This rope design has two elements; an interior core (kern or core) of twisted fibers that supports the central portion of the load on the rope and an outer sheath (mantle) that protects the core also supports a minor part of the load. 1/2" kern mantle rope meets NFPA standards for rescue use.

An example of Kern Mantle Rope

Static Rope is a type of rope designed for low stretching. It's used in rescue, rappelling, and ascending, where stretching would be a disadvantage.

Dynamic Rope is a rope designed for high stretching ability, used to reduce the shock of a falling climber or to be used as an anchoring system. It's usually only used in mountaineering and rock climbing.

Verbal Commands for Rope Use

- Pull: Pull the rope through the rescue system
- Slack: I need some slack because there's too much tension
- Rope: There's too much rope; take up the slack
- Tension: Hold Tight! I'm unstable.
- Rock: Look Out! But don't look up!
- Climb: Means we're ready to go. Are you ready to catch us if we fall?
- On Belay: Means we're ready to catch you if you fall.

The use of ropes for SAR and rescue is a course all unto itself. Many organizations offer courses that include care and inspection of ropes, cleaning, knots, low and high angle rescue usage, and more. It is highly recommended that you complete at least a basic level ropes course before engaging in any SAR activity.

Stokes Basket (Litter)

The Stokes basket is a wire or hard plastic litter designed for use over harsh terrain, on slopes, and in wooded areas to aid in protecting an immobilized patient from further injury during extrication.

Campcraft Students carry out a patient during a first aid class.

Packaging

Packaging refers to the process of securing a patient into a Litter or keeping them in a stabilized position for extrication, be it from a vehicle or pile of debris. Typically taking the form of nylon webbing, packaging secures a patient so that even in the event of the litter being turned upside down, they will not move during transport. Packaging also includes padding areas of the body that may be injured or prone to injury from prolonged immobilization in a litter.

SAR SURVIVAL | 57

Campcraft students demonstrate packaging during a
SAR Responder Class

Preparing a Litter for Transport

- Appoint a Litter Team Boss
- Guide/ Trail clearing personnel should get ahead of the litter team and clear trail and mark paths
- Reserve litter bearers should follow the litter to rotate in to carry when one becomes tired

Lifting the Litter

- A nylon shoulder strap should be attached to the litter and adjusted should the rescuer's hand slip from the litter rail so the patient does not fall
- Litter bearers all face inwards and kneel on knees closest to the subject's feet
- Bearers place both hands on litter rail a comfortable distance apart
- Litter bearers lift with legs and not their back
- Litter Boss gives the command: "Prepare to Lift," and bearers indicate whether or not they are ready
- "Lift" command is given only when all are ready to lift and proceed.
- Lift smoothly without jostling the patient
- Normally, patients are carried feet first toward your travel objective unless going uphill, in which case they will travel headfirst.

Carrying the Litter

- The litter boss gives the instructions to move in an indicated direction
- Litter bearers walk out of step with one another to avoid discomfort
- If terrain becomes too steep for sound footing, the litter should be lowered, and low angle rescue equipment should be employed

- If a litter bearer becomes unsure of their footing or ability to continue, they should speak up and ask the Litter Boss to halt.

7

SAR Knots

A few knots will aid new SAR technicians if they must assist in a rescue or perform patient care that would involve setting up and shelter or giving first aid that may require splinting.

Bowline Knot

The Bowline is a simple knot used to form a fixed loop at the end of a rope. It has the virtues of being easy to tie and untie; most notably, it is easy to untie after being subjected to a load.

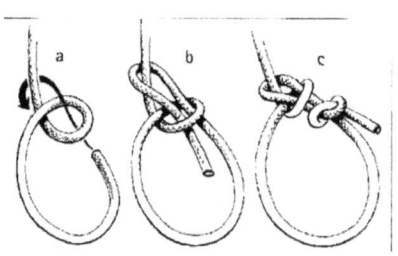

Bowline Knot

Clove Hitch

The clove hitch is two successive half-hitches placed around an object. It is most effectively used as a crossing knot. It can also be used as a binding knot but only when used with an additional safety knot to keep it secure.

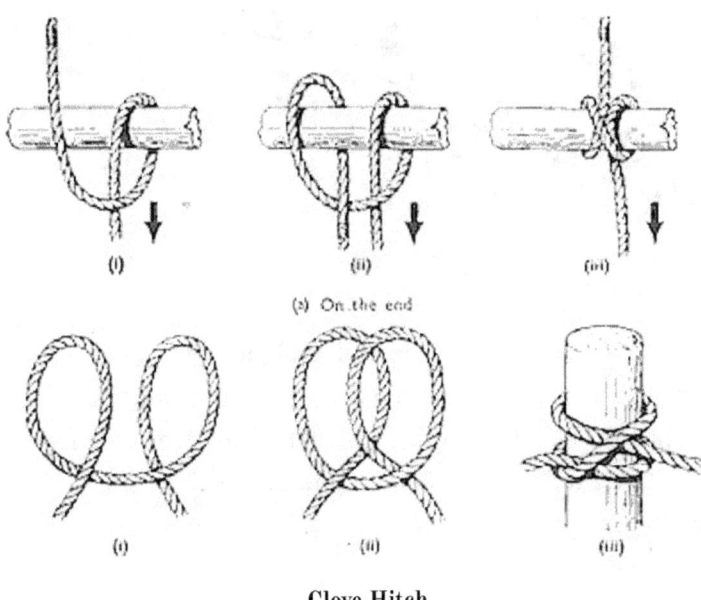

Clove Hitch

Figure Eight Knot

This knot is essential in rescue and rock climbing to stop ropes from running out of retaining devices. Like the overhand knot, which will jam under strain, often requiring the rope to be cut, the figure-of-eight will also stop but is usually more easily undone than the overhand knot.

Overhand Knot

Perhaps the most common and straightforward knot is made by forming a loop and passing a free end around the standing part and through the loop. It's not great for load-bearing as it jams up so tight it must be cut out, but for quick or practical uses, it will do the job.

Sheet Bend Knot

This is simply a knot that joins two ropes together. Doubled effectively binds lines of different diameters or rigidity securely together, although it tends to work loose when not under load.

SAR SURVIVAL | 63

Sheet Bend Knot

8

SAR Response Kit

Equipment Considerations

For SAR purposes, it's recommended that each search member carry a 24hr ready pack, including items for personal use and your lost subject. As a Survival Instructor and a SAR Team Leader, I have come to include the Emergent C's into my ready pack as with proper training; we learn to utilize the Emergent C's for our survival and self-aid needs. With this 12 Category tool system, we're also able to care for patients to some extent by being able to aid them in maintaining their bodies core temperature, treating them for shock, and treating common injuries such as cuts, abrasions, minor burns, swelling, and contusions (bruises).

The Emergent C's (12 C's) are:

1. Cutting tools (Knife, Axe, and Saw)
2. Combustion Device (Ferro Rod)
3. Container (Stainless Steel Bottle/Canteen)

4. Cordage (100ft Bankline or Paracord)
5. Cover (Emergency Survival Blanket)
6. Compass with Sighting Mirror
7. Cloth (3' x 3' 100% cotton material with repair needle)
8. Candling Device (Headlamp with Spare Batteries)
9. Cargo (Roll of Gorilla Tape)
10. Combination Tool (Multi-Tool)
11. Cerate (Medicinal Salve or Small First Aid Kit)
12. Circumvention (Water Filter, snack, mission dependent tools)

We utilize the 12 C's in the following ways for self-aid and patient care:

Cuts and Abrasions

- Water Bottle- Irrigation and wound washing
- Knife & Sail Needle – cleaning of a wound
- Cotton Material- cleaning, bandages & dressings, tourniquet
- Duct Tape- quick Band-Aids, holding bandages in place
- Compass- Magnifying glass for close up inspection of the wound
- Cordage- last resort tourniquet
- Drum liner- bandages, wound cover (occlusive seal)
- Cerate- To provide healing properties and soothing coverage
- Circumvention- Water filters to provide continuous clean water

Breaks, Strains, and Sprains

- Water bottle – used as either a hot or cold pack
- Duct Tape – Used for minor immobilization as well as holding splints in place
- Blanket- used in immobilization, padding between splints
- Drum liners- holding splints in place, improvised slings, water collection
- Cotton Material (bandanna) Slings, Splinting, padding
- Pack- Immobilization

Burns

- Water Bottle- Irrigation, cold pack, stop burning process
- Cotton Material- loosely cover burns, keep moist
- Knife- clean burns
- Needle- clean burns
- Drum Liner- cover burns
- Compass- inspection of burns
- Cerate- Burn relief and healing

Bites and Stings

- Compass- inspection, and location of stingers, splinters, bites, ticks
- Needle- stinger and splinter removal
- Knife- stinger and splinter removal

- Water Bottle- Irrigation, cold compress reduce swelling
- Duct Tape- quick Band-Aids, cover wounds, hold bandages in place, remove stingers
- Cotton material- bandages & dressings, pressure dressings
- Seasonal- Ticks, mosquitoes, snakes, spiders (Blk. Widow & Br. Recluse)
- Cerate- itch relief

Blisters

- Sail needle to lance blister
- Duct tape prevent hot spots cover wound
- Water bottle to clean the wound
- Cerate- wound coverage and protection

Other Uses of the 12 C's

- Water Bottle & cup- preparing medicinal infusions, decoctions, sterilizing items
- Tarp- Stretcher, water catch, protection from sun, wind & rain, splinting material, slings
- Fire Kit- allows heat fights hypothermia, creates fire for cooking medicinals, sterilization, heat compresses, making charcoal
- Compass- self-inspection hard to see areas
- Drum Liners- Improvised Tourniquet to Stop life-threatening bleeding
- Headlamp- allows inspection at night
- Cerate- packing open wounds

- Circumvention- water filtration for ongoing clean water or snacks for prolonged stays

So, whatever else you add into your SAR Ready Pack, I would recommend utilizing the Emergent C's like the foundation of the kit. NASAR recommends the following items and will exceed the national standard when combined with the 12 C's. Where the 12 C's overlap the recommended list, I have omitted the recommendation since we would build off the 12 C's...

Recommended Items

- Sun Glasses/ Goggles/ Eye Protection
- Small Plastic Bags (Sandwich and Freezer Bags)
- Band-Aids of Various Sizes (At least 6)
- Roller Gauze (Kling or Kerlex)
- Mole Skin
- Betadine Prep Pads (2)
- Large Safety Pins (2)
- Wire, small gauge, 12" (snare wire)
- Cotton Swabs (2)
- Benadryl (4)
- Tylenol or Aspirin (4- not given to trauma-related bleeding)
- Ibuprofen (4- can be given to trauma-related bleeding)
- Antacid Tablets (4)
- Razor Blade (1- Single Blade)
- Stormproof Matches (8)
- 55 Gallon Drum Liner (2)

- Candle and Fire Starters (1)
- Betadine Ointments (1)
- Clean Towelette (1)

Additional recommended personal gear

- Adequate Footwear (All-weather/Climbing capable)
- Small Writing Pad and Pencil
- Rain Gear
- Walking stick/ Hiking Staff
- Extra Clothing (Entire Outfit)
- Work Gloves
- Flashlight with batteries
- Insect Repellant
- Trail Snacks or similar food
- Watch that is waterproof and durable

Example of multi-day kit components

For extended searches or multi-day events where inhospitable weather is expected where you are unfamiliar with the territory, it is recommended that you also carry:

- Stiff Brimmed Hat
- Rain Cover for Backpack
- Extra Socks
- Tweezers
- Bouillon Cubes
- Bush Pot
- Water Purification Device (Filters or Tablets)
- Sleeping Pad
- Non-Perishable Food
- Toilet paper
- Plastic Tarp
- Sunscreen
- Flagging Tape

Plus, personal medications you require. Keep in mind that you will not use all or even half of the equipment you carry on any single mission; the ready pack is kept stocked so that you are always prepared to go at a moment's notice.

9

SAR Survival

Survival for SAR purposes means the ability to endure brutal conditions (bad weather, injury, etc.) for periods up to 72 hours with only the equipment on you or that you can improvise from surrounding materials off the landscape. When thinking of survival as it relates to SAR, we look at the survival triangle:

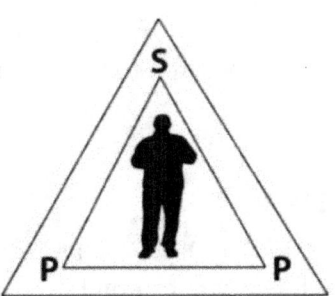

S- Spiritual Care and Emotional Support

Spiritual and Emotional Support goes hand in hand with the standard psychological support you would provide a patient suffering from shock. Should you encounter a subject who becomes your patient for a period, remain positive, connect with them on a spiritual level and keep them talking. This goes a long way in changing their outlook and motivating them to keep going despite a severe injury. Talk about faith, family, or anything else that fuels their mind to press on.

P- Physical Ability

Do you or does your patient have the physical capacity to rescue themselves? If not, what are your limitations, and what can you do to remedy them in the short term while awaiting additional support. From here, we move into the acronym S.T.O.P.

> S- Sit and Stop what you are doing
> T- Think about your circumstances
> O- Observe your surroundings
> P- Plan a way to resolve your problem

Once you STOP, you move on to the final side of the Survival Triangle:

P- Practical Application

Is the plan you and perhaps your patient appears to be actionable? E you confident you can pull it off without the risk of additional injury, etc., and could you continue to survive if other resources were delayed for any reason.

When one side of the triangle fails, your overall planning fails, and when too much focus is given to one area, the other areas falter, leading to more adverse conditions. Improvise-

Adapt- Overcome; this motto applies to survival, especially about SAR, where care for you, the searcher, and the patient is of paramount priority. If you, the searcher, becomes injured or incapacitated, you put a strain on SAR resources and prolong the potential rescue of the lost subject(s). So make every effort to care for yourself properly so that the emergency remains the lost subjects and not your own.

The Rule of 3's

The rule of 3's is a simple concept that will aid you in prioritizing during a survival or emergency.

- 3 Minutes: The brain can, for the most part, go for three minutes deprived of oxygen before damage begins to occur.
- 3 Hours: The human body can only withstand three hours of exposure to extreme elements.
- 3 Days: The human body can go for three days without water.
- 3 Weeks: The human body can only go an average of three weeks without food before beginning to shut down critical body systems.

These are essential facts to remember when prioritizing your plan of action. This will assist you in becoming comfortable in an uncomfortable situation.

The 3 Killers

- Hypothermia: The condition at which an organism's temperature drops too far below that required for normal metabolic function. When a body is exposed to cold, its internal mechanisms may be unable to replenish the heat lost to the body's surroundings.
- Hyperthermia: Also known as "Heat Stroke," is an acute condition that occurs when the body absorbs more heat than it can dissipate. Heat regulating mechanisms shut down, and body temperature climbs uncontrollably.
- Shock is a severe and life-threatening condition where insufficient blood flow reaches body tissues. Because blood carries oxygen and nutrients around the body, this reduced blood flow causes tissues and organs to cease their normal function. This is known as hypoperfusion or an inadequate exchange of blood and oxygen between body tissue.

Survival's Base 5: Survival Priorities

By utilizing STOP, the Survival Triangle, and understanding the 3 Killers and Rule of 3's, the SAR Responder will be equipped to prioritize the five most important areas of need efficiently and effectively.

Base 5 is what humans need to survive in any given situation, including day-to-day life. Failure to address one of these areas will result in emotional, physical, and mental instability and other health conditions in more severe cases.

The Base 5 appear below in their prioritized order:

1. Shelter (1 and 2 can be changed to suit environmental conditions)
2. Fire
3. Water
4. Security/ Safety
5. Food

Shelter

Assuming we have adequate oxygen, we know that according to the Rule of 3's, the average person can only withstand 3 hours of exposure to extreme elements before becoming ill or injured. Harsh elements could be anything from the hot sun to the wind and rain of a crisp fall day. Shelter becomes #1 whenever the weather is potentially not in your favor and will provide a sanctuary to you and your patient.

In the event rescue or extrication is delayed, throwing up a hasty shelter, which we call the 5 Minute Shelter, is a great idea that will enable you to comfort your patient and provide them some element of protection from the elements.

Fire

Fire is #2 and ties in with the shelter. We rate it as the second most important on our list because there are circumstances when a fire is not necessary for survival where a cover would provide additional benefits. However, if you are in the backwoods of Kentucky on a search mission, a fire will most certainly be your number one priority. It will not only enable you to rewarm yourself and your patient but provide warming fluids to a potentially hypothermic patient. So, carrying several ways to make a fire is a must.

Water

Again, when looking at the Rule of 3's, we know that we can only survive about three days without water. Obtaining clean, potable water is a skill everyone should learn, even if they are not outdoor-minded people. Water will not only allow you to hydrate, but cool burns irrigate wounds, and begin rewarming frost nipped extremities.

Security and Safety

This may come as a surprise, but if you don't feel relatively safe where you are, you will not be able to think clearly and prioritize effectively for longer-term survival. If you are in an area with known threats to you, the rescuer, or the patient, you may need to make an effort to move your location or bolster security for the sake of safety. For example, if you find your patient next to an active bear den, you may want to move them away, or if for some reason you cannot, bolster your camp security by creating a large fire, debris field to slow an approaching bear, make a weapon, etc.

Food

Food is the lowest on our list of priorities because it's the one thing we can go the longest without. As such, it's unlikely we will require food in the SAR scenario unless it's to give a supplementary snack to a patient or to fuel the body of a rescuer. Carbohydrate-rich snacks are beneficial over proteins which require additional water to digest. For a potentially hypothermic patient, a snack rich in carbs will stoke their inner fire to create warmth due to the metabolic rate increase.

Skills Important to SAR Responders

The three most essential survival skills for SAR responders are those taught in our Basic Survival courses. These skills are the ability to make a fire in five minutes or less. The ability to establish a shelter also in five minutes or less and boil water for the sake of making it potable again in five minutes. So these three skills must be completed within 15 total minutes. We strive to do all three in less than 15 minutes because as we begin to get cold and remain out in the elements, we start to lose manual dexterity when we become hypothermic or hyperthermic. The loss of manual dexterity makes it very difficult to utilize tools to affect survival. When another patient's life may depend on our ability to use our tools, we must come prepared.

Fire starting is a vital SAR Skill

You should always include at least two Bic lighters for your fire staring methods. They should always serve as your first fire-making device, but if they become lost or damaged, a ferrocerium rod is the best choice as a backup. In addition to the fire-making device, a waterproof fire starter or any other similar product is a must-have. The simple cotton ball and vaseline is NOT acceptable for a SAR Kit as they are not waterproof and can and do fail when they are most needed.

When striking your Ferro rod, use a striking device or the spine of your knife when at all possible. Do not use your cutting edge; not only is it dangerous, but it degrades your knife-edge, which may be needed for other essential tasks.

When boiling water from a stream or river, cook in the cup while drinking from the bottle. The cup will boil and cool faster than an entire bottle's worth of water, so as the cup's worth is cooling in your bottle, you can start a new batch boiling. It is only necessary to bring water to a rolling boil under 5,000 feet in elevation to be potable; add 30 seconds for every 1000 feet in height above 5,000 feet to be safe because water begins to boil at a lower temperature above 5000 feet. Typically, there is no need to filter it unless it's full of sediment or potentially contaminated by chemicals.

Shelter

The Hasty Shelter or 5 Minute Shelter is nothing more than a simple Lean To created with an Emergency Survival Blanket. This shelter style is sufficient to keep wind and rain at bay for short periods and helps recycle the radiant heat. When placed directly to the skin, the blankets themselves can reflect up to 90% of the body's heat back to itself, which goes a long way when rewarming a hypothermic patient. Conversely, on a sunny day where your patient requires shade, the reflective side makes a great sun reflector which also reflects some heat away from the patient while also serving as an impromptu signaling device.

Night Exercise

With most Basic SAR courses, the final exam where students apply what they have learned in a mock exercise is completed at night. In a simulated search, you can expect to experience much of what would happen during an actual SAR Mission. You will see Incident Command Established and experience a briefing, receive a radio, be given a search area, and be expected to report clues. Upon finding your lost subject, you will need to care for or perform the appropriate SAR tactic to ensure their safety and eventual transfer of care to EMS or Law Enforcement. It's an exciting exercise meant to provide you with realistic expectations in a controlled environment.

The author's 13yr old son training at night in SAR Survival

The following are some links of interest for additional information on SAR Training and Survival training:

Campcraft Outdoors- Kentucky
www.campcraftoutdoors.com

National Association of Search and Rescue
www.nasar.org

Mountain Rescue Association
www.mra.org

Emergency Response International
www.eri-online.com

10

SAR Documents

The following are an example of documents used within SAR Teams. These particular documents are those we utilize in our local department and may be specific to Kentucky SAR Teams.

Still, much of the information is in a standardized form, which may be helpful to you for education or creating something similar for your own SAR Team.

That other may live...

TEAM ASSIGNMENT	INCIDENT NAME		OPERATIONAL PERIOD	TASK NUMBER
SEGMENT DESIGNATION	SEGMENT AREA SQ. MI.	TIME TO SEARCH HRS	SEARCHERS ASSIGNED	SEARCHER SPACING FT
RESOURCE OR SEARCH TYPE				# OF TRACKS

PERSONNEL ASSIGNED (* IF CHECKED (√) INDICATES A MEDICAL CONCERN) PLEASE RECORD TEAM LEADER ON LINE 1

*	NAME	AGENCY	*	NAME	AGENCY
1			7		
2			8		
3			9		
4			10		
5			11		
6					☐ ADDITIONAL NAMES ATTACHED

ASSIGNMENT

☐ MAP(S) ATTACHED

PREVIOUS AND PRESENT SEARCH EFFORTS IN AREA

☐ DEBRIEFING INFO ATTACHED

DROP OFF AND PICKUP INSTRUCTIONS

EQUIPMENT ISSUED

TEAM COMMUNICATIONS PLAN

RADIO CALL / TEAM DESIGNATION:		COMMAND POST CELL #:	
FUNCTION	FREQUENCY	CHANNEL DESCRIPTION	CHANNEL
SEARCH (TEAM ⇔ BASE)			
CONFINEMENT (TEAM ⇔ BASE)			
TACTICAL (TEAM ⇔ TEAM)			
PREPARED BY		DATE PREPARED	TIME PREPARED

SAR 104 NICOLET SEARCH TEAM 8/22/07	COPIES TO	☐ PLANNING ☐ COMMUNICATIONS ☐ OPERATIONS ☐ TEAM	BRIEFER	
			TIME OUT	TIME RETURNED

SIDE 1

CLUE LOG		1. INCIDENT NAME		2. DATE	3. INCIDENT NUMBER	
CLUE #	ITEM FOUND		TEAM	DATE/TIME	LOCATION OF FIND	INITIALS

SAR 134
BASARC 3/98

CLUE REPORT	1. INCIDENT NAME		2. DATE	3. INCIDENT NUMBER
4 CLUE NUMBER	5. DATE/TIME LOCATED		6. TEAM THAT LOCATED CLUE	

7. NAME OF INDIVIDUAL THAT LOCATED CLUE

8. DESCRIPTION OF CLUE

9. LOCATION FOUND

10. TO INVESTIGATIONS
- [ ] **URGENT REPLY NEEDED**, TEAM STANDING BY TIME _____
- [ ] INFORMATION ONLY

11. INSTRUCTIONS TO TEAM
- [ ] COLLECT
- [ ] MARK AND LEAVE
- [ ] DISREGARD
- [ ] OTHER _____

CLUE & SEGMENT PROBABILITIES TO BE COMPLETED BY PLANS

12. CLUE PROBABILITY	13. SEGMENT PROBABILITY	LIST SEGMENTS
[ ] VERY LIKELY A GOOD CLUE	VIRTUALLY 100% CERTAIN CLUE MEANS SUBJECT IS IN THESE SEGMENTS	
[ ] PROBABLY A GOOD CLUE	VERY STRONG CHANCE THAT CLUE MEANS SUBJECT IS IN THESE SEGMENTS	
[ ] MAY BE A GOOD CLUE	STRONG CHANCE THAT CLUE MEANS SUBJECT IS IN THESE SEGMENTS	
[ ] PROBABLY NOT A GOOD CLUE	BETTER THAN EVEN CHANCE THAT CLUE MEANS SUBJECT IS IN THESE SEGMENTS	
[ ] VERY LIKELY NOT A GOOD CLUE	NO INFORMATION FROM THE CLUE TO SUGGEST SUBJECT IS OR **IS NOT** IN THESE SEGMENTS	
[ ] DON'T KNOW	BETTER THAN EVEN CHANCE THAT CLUE MEANS SUBJECT IS **NOT** IN THESE SEGMENTS	

COPIES
- [ ] PLANS   [ ] ATTACH TO CLUE
- [ ] INVESTIGATIONS   [ ] OTHER
- [ ] DEBRIEFING _____

	STRONG CHANCE THAT CLUE MEANS SUBJECT IS **NOT** IN THESE SEGMENTS	
VERY STRONG CHANCE THAT CLUE MEANS SUBJECT IS **NOT** IN THESE SEGMENTS	
VIRTUALLY 100% CERTAIN CLUE MEANS SUBJECT IS **NOT** IN THESE SEGMENTS	

SAR 135 BASARC 3/98 | **14. PREPARED BY** | **15. CLUE & SEGMENT PROBABLITIES PREPARED BY**

SUBJECT INFORMATION						
NAME						AGE	SEX
NAME TO CALL			EXPECTED RESPONSE			
SUBJECTS PLANS OR INTENT							
PHYSICAL DESCRIPTION			CLOTHING DESCRIPTION			
HEIGHT	WEIGHT	BUILD				
RACE	COMPLEXTION					
EYES	HAIR					

<u>GENERAL INFORMATION ABOUT SEARCH PROCEDURES</u>

1. **TEAM BRIEFING TO INCLUDE:**

   A. THAT ALL TEAM MEMBERS ARE THOSE ASSIGNED TO THIS TEAM. DO NOT LET ANY OTHERS JOIN WITHOUT THE OPERATIONS SECTION PERMISSION. MAKE SURE THAT ALL MEMBERS ARE SIGNED IN WITH COMMAND.
   B. PROPER CLOTHING FOR WEATHER AND TERRAIN, INCLUDING_____
   C. STURDY BOOTS / SHOES. TEAM LEADERS, YOU HAVE THE RIGHT NOT ALLOW TENNIS SHOES OR OTHER SUCH FOOTWEAR TO BE WORN.
   D. 2 WATER BOTTLES AND/OR CANTEENS OF AT LEAST 1 LITER EACH.
   E. EACH TEAM MEMBER HAS A READY PACK (USE SAR FORM SAR 10 AS A GUIDE) AND A PERSONAL FIRST AID KIT (INCLUDING BEE STING KITS IF ALLERGIC TO STINGS).
   F. NO ONE UNDER THE AGE OF 16 IS ALLOWED ON YOUR TEAM. 16 & 17 YEAR OLDS MAY, AT YOUR DISCRETION, BUT ONLY WITH THEIR PARENT AND/OR GAURDIAN ALSO WITH YOUR TEAM.
   G. BRIEF THE TEAM AS TO ANY HAZARDS THEY MIGHT ENCOUNTER IN YOUR SEARCH SEGMENT.
   H. BRIEF THE TEAM AS TO HOW LONG THEY ARE PROJECTED TO BE IN THE FIELD.

2. **ACTIVITIES LOG:**

   A. AS TEAM LEADER, YOU MUST MAINTAIN A TEAM ACTIVITIES LOG OF YOUR ASSIGNMENT, TO BE TURNED IN TO THE PLANNING SECTION UPON COMPLETION OF YOUR TASK ASSIGNMENT.
   B. RECORD ALL RELEVANT ACTIVITES OF YOUR TEAM; SEGMENT ASSIGNMENT, START AND ENDING TIMES, AREAS YOU SEARCHED, AREA YOU WERE UNABLE TO SEARCH, HAZARDS IN THE AREA, OTHER PROBLEMS ENCOUNTERED, SUGGESTIONS FOR A RE-SEARCH OF THE AREA, AND ANYTHING ELSE THAT YOU MIGHT FEEL IS IMPORTANT TO THE PLANNING SECTION OR THE INVESTIGATORS.
   C. **IMPORTANT** – RECORD SEARCHER SPACING THAT CHANGED,THE REASONS FOR THE CHANGE AND THE EXACT LOCATIONS ON THE SEGMENT MAP THAT THE CHANGE OCCURED.

3. **HANDLING CLUES:**

   UPON THE DISCOVERY OF A POSSIBLE CLUE, THE TEAM WILL STOP AND THE TEAM LEADER WILL EXPLORE THE CLUE. IF DEEMED A RECENT OR RELEVANT CLUE, THE TEAM LEADER WILL CONTACT THE COMMAND POST IMMEDIATELY AND BE PREPARED TO RELAY THE FOLLOWING INFORMATION:

   A. TEAM THAT HAS LOCATED THE CLUE.
   B. TIME LOCATED.
   C. NAME OF INDIVIDUAL WHO LOCATED THE CLUE.
   D. GPS COORDINATES OF CLUE LOCATION, INCLUDING SEARCH SEGMENT.
   E. BRIEF DESCRIPTION OF CLUE.
   F. BRIEF DISCRIPTION OF LOCATION FOUND.
   G. STANDBY FOR FURTHER INSTRUCTIONS FROM COMMAND.
   H. IF DEEMED A RELEVANT CLUE BY COMMAND, THE TEAM LEADER WILL NEED TO COMPLETE A CLUE REPORT DURING DEBRIEFING.

4. **DEBRIEFING:**

   A. UPON TASK COMPLETION, YOU MUST RETURN TO THE DEBRIEFING AREA AND DEBRIEF YOUR ASSIGNMENT WITH THE PLANNING SECTION BEFORE DOING ANYTHING ELSE.
   B. NO TEAM MEMBERS ARE TO BE RELEASED TO STAGING OR REHAB UNTIL YOUR TEAM HAS BEEN DEBRIEFED AND RELEASED FROM YOUR ASSIGNMENT.
   C. IF MEMBERS OF YOUR TEAM ARE GOING TO BE LEAVING THE SEARCH AREA, HEADING HOME, ETC.. THEY MUST SIGN OUT WITH COMMAND.

IMPORTANT….IMPORTANT…..IMPORTANT - DURING THE SEARCH AND SEARCH DEBRIEFING, WE ARE RELYING ON YOUR HONESTY TO TELL US EXACTLY WHAT YOU ENCOUNTERED. IF YOU DID NOT SEARCH AN AREA, TELL US, IF YOU SEARCHED MORE THAN YOUR AREA, TELL US. IT MAKES A DIFFERENCE TO THE PLANNING FOLKS….

REMEMBER...

...THESE THINGS WE DO, SO THAT OTHERS MAY LIVE...

SAMPLE SEARCH TEAM MINIMUM PERSONAL EQUIPMENT LIST

The following equipment is commonly compiled to form what is referred to as a "24-hour ready pack". Such a pack holds those items that would allow the holder to function in a safe, effective, and efficient manner during a SAR incident. Some items may be carried on a belt, in pockets, or strapped to the person. This equipment should be carried on all missions in rural or wilderness areas and is considered the minimum amount

PERSONAL FIRST AID KIT AND SURVIVAL KIT

√	QTY	DESCRIPTION	√	QTY	DESCRIPTION
	10	Acetaminophen or aspirin tablets		4	Roller Gauze Bandage
	10	Antacid tablets		4	Safety pins, large
	10	Antihistamine, 25mg Benadryl		1	Splinter forceps, tweezers
	6	Antiseptic cleansing pads		1	Space type blanket or space type sleeping bag
	1	Antiseptic ointment (tube)		2	Towelette, clean
	6	Band aids, various sizes		10	Water Purification tabs sealed or water purification device
	4	Cotton swabs, non sterile		10	Antihistamine, 25mg Benadryl
	1	Duct tape, 5-10 ft		2	Nitrile Gloves (Latex should be avoided)
	4	Leaf bag, large			
	16	Matches in a waterproof container		*2	Quarters for phone call (SARTEC)
	1	Moleskin (at least 3x4")		*1	Razor blade, single edge safety type (SARTEC)
	1	Plastic bag, zip lock, qt. size, for kit		*1	Candle, long burning (SARTEC)

PERSONAL SAR EQUIPMENT

√	QTY	DESCRIPTION	√	QTY	DESCRIPTION
	4	Bags, various sizes, zip locked		1	Pack, 1800 cubic inch (minimum)
	1	Bandanna or handkerchief		1	Pad and pencil
	1	Cap or other wide brimmed headgear		1	Rainwear, durable
	1	Clothes bag, waterproof		1	SAR personal identification
	1	Clothing, adequate for climate		1	Scissors, multi-purpose
	1	Clothing, extra set, suitable for climate		1	Socks, extra pair
	1	Compass, orienteering		1	Sunscreen lotion
	2	Flagging tape, roll		1	Tissue papers or baby wipes
	1	Flashlight or lantern		1	Tracking stick, minimum of 42" long
	1	Flashlight extra, extra batteries and Bulb		1	Watch
	1	Footwear, sturdy, adequate for climate		2	Water containers, at least liter size
	1	Gloves, durable, even in summer		1	Watch
	1	Goggles, or eye protection, clear		2	Water containers, at least liter size
	1	Insect repellent		8	Wire ties, elastic, self locking
	1	Knife, multi-purpose		1	Whistle
	1	Lip balm, with sunscreen		2	ICS 214 Unit Log Form
	1	Measuring device, 18 in. minimum		1	Lost Person Interview Form
	1	Mirror, small		1	Grid Reader (UTM)
	1	Nylon twine or small rope, 50 feet		1	Pace Counter
	*1	Sterno or stove (SARTEC)			
	*1	Safety Rope 75ft (one rescuer lifeline/NFPA) (SARTEC)			Nicolet Search Team Shirts / Uniform clothes
	*1	Shelter Material, 8x10 plastic or coated nylon (SARTEC)			Radio Battery Charger
	*2	Prusik slings (suitable for 9mm to 11 mm rope) (SARTEC)			Set of flashlight batteries
	*2	Carabiners (locking) (SARTEC)			Personal Hygiene
	*1	Webbing, 1" tubular - length suitable for harness (SARTEC)			Sleeping Bag
	*1	Wire, 5-10 ft, woven steel (SARTEC)			Gaiters
	*1	Metal cup or pot (SARTEC)			Trail snacks

SAR 22
SAMPLE SEARCH TEAM
01/05/2022

Items marked are required only if you are attempting the SARTEC I or II practical field test.
Items in RED are updates 1-16-2009

www.ingramcontent.com/pod-product-compliance
Lightning Source LLC
Chambersburg PA
CBHW071833290426
44109CB00017B/1816